SIMPLE MACHINES ALL AROUND US

WEDGES AT WORK

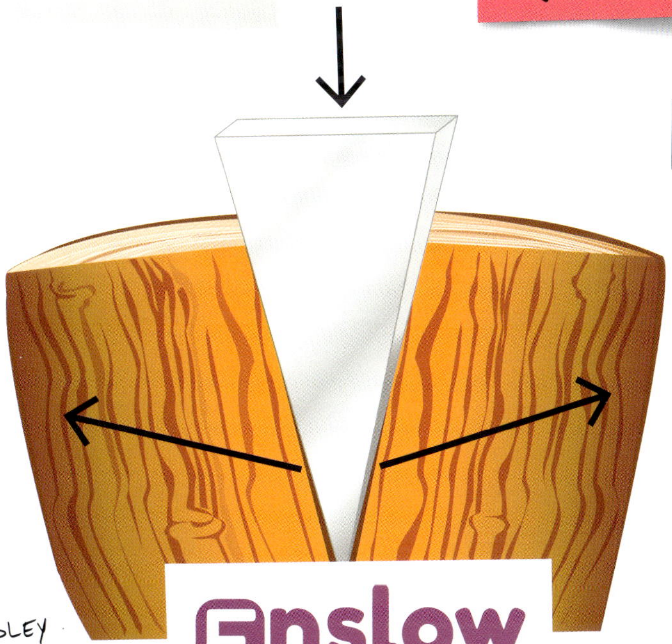

By Maggie S. Wesley

Enslow Publishing

DISCOVER!

Please visit our website, www.enslow.com. For a free color catalog of all our high-quality books, call toll free 1-800-398-2504 or fax 1-877-980-4454.

Library of Congress Cataloging-in-Publication Data

Names: Wesley, Maggie S., author.
Title: Wedges at work / Maggie S. Wesley.
Description: New York : Enslow Publishing, [2022] | Series: Simple machines all around us | Includes index. | Audience: Ages 6 | Audience: Grades K-1
Identifiers: LCCN 2020028567 (print) | LCCN 2020028568 (ebook) | ISBN 9781978520967 (library binding) | ISBN 9781978520943 (paperback) | ISBN 9781978520950 (set) | ISBN 9781978520974 (ebook)
Subjects: LCSH: Wedges–Juvenile literature.
Classification: LCC TJ1201.W44 W47 2022 (print) | LCC TJ1201.W44 (ebook) | DDC 621.8–dc23
LC record available at https://lccn.loc.gov/2020028567
LC ebook record available at https://lccn.loc.gov/2020028568

Published in 2022 by
Enslow Publishing
101 West 23rd Street, Suite #240
New York, NY 10011

Designer: Katelyn E. Reynolds
Editor: Kristen Nelson

Photo credits: Cover, pp. 1 ttsz/ iStock / Getty Images Plus; cover, pp. 1–24 (background and typeface) Reytr/Shutterstock.com; cover, pp. 1–24 (paper) jannoon028/Shutterstock.com; cover, pp. 1–24 (note sticker) Kindlena/Shutterstock.com; p. 5 VectorMine/ iStock / Getty Images Plus; p. 7 BRIAN MITCHELL/ Corbis Documentary/Getty Images; p. 9 Connor Walberg/ The Image Bank / Getty Images Plus; p. 11 robcruse/E+/Getty Images; p. 12 Marlene Bocast / EyeEm/Getty Images; p. 13 Jan Woitas/picture alliance via Getty Images; p. 15 Malorny/Moment/Getty Images; p. 17 Werner Forman/Universal Images Group/Getty Images; p. 19 (inset) Creti Stefano/Moment/Getty Images; p. 19 photominus/ iStock / Getty Images Plus; p. 21 (doorstop) Louella938/Shutterstock.com; p. 21 (knife) eskymaks/ iStock / Getty Images Plus; p. 21 (ax) SolidMaks/ iStock / Getty Images Plus; p. 21 (shovel) ElephantCastle/ iStock / Getty Images Plus; p. 21(nail) Narongsak Kumma / EyeEm/Getty Images.

Portions of this work were originally authored by Gillian Houghton Gosman and published as *Wedges in Action*. All new material this edition authored by Maggie S. Wesley.

Printed in the United States of America

CPSIA compliance information: Batch #CSENS22: For further information contact Enslow Publishing, New York, New York, at 1-800-398-2504.

Contents

Boldface words appear in Words to Know.

What Is a Wedge?

A wedge is a **triangular** tool, commonly made of metal, wood, or plastic. One end is wide. The other end comes to a point or a sharp edge. Wedges are simple machines. Simple machines make work easier. They have no or few moving parts.

WEDGES ARE ONE OF SIX SIMPLE MACHINES.

Wedges vs. Inclined Planes

Wedges and inclined planes are simple machines that look alike. An inclined plane is a ramp that is sloped. It's used to move loads more easily. But inclined planes don't have to move to do work. Wedges, though, must be moved to do work.

INCLINED PLANE

How It Works

A wedge works by changing the direction of the **force** applied to it. For example, an ax is swung down toward a piece of wood. The force of the swing is directed down. The blade, or thin edge of the wedge, cuts the wood by directing the force sideways into the wood.

AN AX IS A DOUBLE WEDGE, OR TWO INCLINED PLANES PUT TOGETHER.

A rubber doorstop is another example of a wedge. When you slide a doorstop under a door, the force of your action travels in line with the floor. The wedge's shape and the **friction** created by it moving against the floor changes the direction of the force and keeps the door in place.

A DOORSTOP IS A SINGLE WEDGE.

A Wedge's Jobs

A wedge does many different jobs. It can be used to cut through **materials**, such as a knife does. Wedges can also be used to dig or **scrape**. They may be used to force or hold something open. A wedge, such as a nail, can hold two things together too.

BOTH A KNIFE AND YOUR TEETH ARE WEDGES THAT HAVE A SHARP EDGE USED TO CUT THROUGH FOOD.

The Advantage

The job of a simple machine like a wedge is to make work easier. The help that the machine gives is called its mechanical advantage. The longer and thinner a wedge is, the greater its mechanical advantage. This means it takes less effort to do the work.

MANY WEDGES HAVE A HANDLE. THIS MAKES THEM EASIER TO USE.

Wedges and People

People have used wedges for millions of years! Early people living in Africa around 2.6 million years ago used stone hand axes to cut and split food and tree branches. Native Americans also used wedges made from animal bone to cut wood used to build their boats and homes.

NATIVE AMERICANS USED WEAPONS CALLED TOMAHAWKS. THEIR BLADES WERE WEDGES!

Wedges are used today too. Woodworkers and stonemasons use wedges called chisels. Wedges are also used in farming and building. Earth-moving machines, such as backhoes and plows, use wedged scoops to cut into and move the ground. A simple shovel does the same work.

CHISEL

Find Wedges Around You

Wedges are an important tool for everyone! Your knife, fork, and spoon can each be used as a wedge, scooping or cutting your food into bite-sized servings. A sandbox shovel is used to dig and lift sand. Look around your home. How many wedges can you find?

DOORSTOP

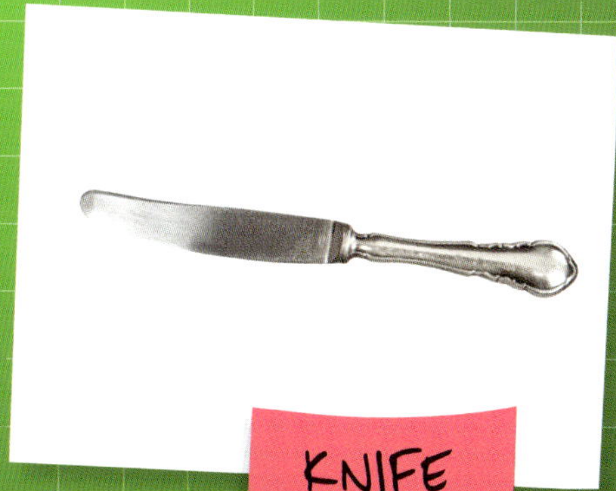

KNIFE

SHOVEL

AX

NAIL

Words to Know

force A power or effect that is able to change the speed or direction of something.

friction The force that slows motion between two objects touching each other.

material Matter from which something is made.

scrape To remove by repeated strokes with something sharp or rough.

triangular Shaped like a triangle.

For More Information

Books

Blevins, Wiley. *Let's Find Wedges*. North Mankato, MN: Pebble, an imprint of Capstone, 2020.

Crane, Cody. *Simple Machines*. North Mankato, MN: Children's Press, 2019.

Websites

Facts About Simple Machines

easyscienceforkids.com/all-about-simple-machines/
Get to know simple machines here!

Simple Machines Facts

www.coolkidfacts.com/simple-machines-facts/
Find out more examples of all kinds of simple machines here.

Publisher's note to educators and parents: Our editors have carefully reviewed these websites to ensure that they are suitable for students. Many websites change frequently, however, and we cannot guarantee that a site's future contents will continue to meet our high standards of quality and educational value. Be advised that students should be closely supervised whenever they access the internet.

Index